Assignment Earth: 2012

The End Or
A New Beginning?

Written by
Michela Hawkins

DEDICATION

This book is dedicated to all the open minded people of this world. Those people who can look beyond the boundaries of our universe and not have their ideas, questions and thoughts be confined by it. This book is for those people, thanks to the media made available to us and the scientists who want to share with us the knowledge of what they have found of the many ancient civilizations that have long since come and gone from our world, who want to see.

This is a book dedicated to honor the devotion of those many scientists, archeologists, paleontologists, seekers of the truth and explorers. To all those who have dared to question who we are, where we came from, and what our future might have in store for us. And to those who have spent endless hours trying to make sense out of what they discover and have the courage to pursue this information and present it to us for examination.

And, also, to the great oracles, prognosticators and seers with the courage to make these predictions and dare to challenge us to do something about what they have predicted.

INTRODUCTION

Are we so arrogant and naïve to believe that we are alone in this universe and beyond? Do we really think all those missing periods in time that leave us with all those unanswered questions of our past can't be explained by a totally different scenario than what our religious backgrounds and foundations of our upbringings have taught us to believe?

There have been numerous new discoveries made possible by our current technologies that are available for our perusal. Almost every week there is something found and reported in the news that challenges our old beliefs. Some new star or planet is found that previously couldn't be seen. Some lost civilization is uncovered in areas that were never explored before. The lost civilizations found to have advancements far beyond what they should have had for that period in time. How can that be explained?

Why did the Mayans calculate the end of their calendar as December 21, 2012? Were the predictions of Nostradamus just the ranting of a mad man? How is it that American Indian tribes speak of the gods from the stars and even have ancient drawings on their cave dwellings that resemble something from outer space ?

Why did the Pope in May of 2008 make an official announcement that just because there may be other life out there, it would still not conflict or contradict all of our religious beliefs?

What about Area 51? Was this a hoax or a cover up by our government, the government we can't exactly trust to tell us the truth. How can we explain so many eye witnesses who are now coming forward to report what they know?

Can you ignore the mass sightings of strange objects in our airspace? Would our astronauts lie about their sightings? Can we believe them?

Let me challenge you to read what I have to say and then judge if it is worthy of some re-thinking on your part or at least a consideration to do so.

In no way will I ever give up my faith and belief in God. This book does not diminish or replace my Christian belief. I believe that science and religion can coexist without eliminating our personal beliefs. This is an attempt to share information I have gathered. You can make up your own mind.

TABLE OF CONTENTS

Chapter One:
THE WORLD'S GREATEST SEER?
Page 1...6

Chapter Two:
THE MAYANS AND THEIR CALENDAR
Page 7...12

Chapter Three:
THE HOPI INDIANS OF ARIZONA
Page 13...16

Chapter Four:
OTHER ORACLES AND PROPHETS
Page 17...20

Chapter Five:
THE SUMERIANS AND THE EGYPTIANS
Page 21...34

Chapter Six:
THE SKULLS OF THE ANCIENTS
Page 35...40

Chapter Seven:
STONE MONUMENTS OF THE WORLD
Page 41...46

Chapter Eight:
ANOMOLIES LEFT BEHIND
Page 47...54

Chapter Nine:
THE BOOK OF REVELATION
Page 55...58

Chapter Ten:
BRIGHT LIGHTS IN THE SKIES
Page 59...66

Chapter Eleven:
ABDUCTIONS AND MISSING TIME
Page 67...70

Chapter Twelve:
ARE WE TAKING 2012 SERIOUSLY?
Page 71...74

Chapter Thirteen:
ASSIGNMENT: COLONIZATION
Page 75...80

Chapter Fourteen:
WHAT DOES ALL THIS MEAN?
Page 81...84

Chapter One

THE WORLD'S GREATEST SEER?

Why have we always been so consumed with a thirst for knowledge of what our future may bring? Why have the stars and planets played such a vital role in every civilization from the beginning of our recorded history? Is it so that we might be able to affect the changes we need to employ to avoid the catastrophes and events predicted by the sages?

When we think of famous prognosticators, the first one that probably is the most familiar to us is Michel de Nostradamus in the mid fifteen hundreds. His quatrains have been the center of controversy for centuries. They were written in several languages, obscure symbols, and vague references were used to keep them from being viewed as appearing to be blasphemy to the church. They have been studied and analyzed by some of the finest minds and scholars throughout the ages.

Recently in 1994, a discovery of yet another book was found that had been in the Vatican library, but locked away for centuries. How and why it was held in secret for so many years is not sure. Some say that if it had been revealed earlier that it maybe would have presented a challenge to our religious beliefs as we know them. The "Lost Book of Nostradamus", which has not yet been totally accredited to Michel, is full of rather primitive illustrations of symbolism representing our past and leading us to our future. Although Michel Nostradamus was not said to be artistically endowed, one of his sons, Cesar, made his living painting portraits. It is by virtue of one of Cesar's paintings of his father that we have a concept of what Michel may have looked like. Michel died when Cesar was only eleven.

Many hours were dedicated to writing the original quatrains and often Cesar would be leaning over the shoulder of his Father watching him toil away. It's not known for sure if Michel actually was sharing his knowledge with Cesar or if Cesar was hoping to steal his father's attention by hanging around while he wrote in the chance they would play together soon. There was some indication that Michel did not wish his talent for predicting on anyone. This ability placed a huge burden on him to constantly produce accurate information. Somewhere in one of the writings he said that "it", meaning the burden of being a seer, ended with him.

The controversy associated with the quatrains begins with the translation of the writings themselves.

In those days, one would have to be extremely careful not to write anything that would be contrary to the teachings of the times for fear of reprisal. Michel de Nostradamus wrote utilizing a sort of code through the rhymes in the quatrains. It's this interpretation of the symbolism in the rhymes that makes it difficult to know exactly what he meant by them. However, some of the prophet's wording is abundantly clear in its meaning.

We are all aware that many of his predictions have come true. Out of the thousand or so predictions he made half of them have come to pass. What we are afraid of is the apocalypse forecast for our future. If he was so right about his other visions, is he going to be right now? Are we entering the end of times now? Will his prediction of a great cataclysmic event happening in December 21, 2012 come true? And what of the plagues, famine, war, and arrival of the Anti-Christ and return of Christ leading up to it? Aren't we seeing some of these now? Will it all come to pass as he predicted it will?

What cataclysmic event was he forecasting? Keep in mind that Nostradamus was very well versed and learned in astronomy, astrology, alchemy math, and chemistry. He predicted eclipses and solar events thousands of years before they would occur. The cataclysmic event he refers to is happening on the winter solstice, December 21, 2012. There is a unique galactic alignment that happens only once every 26,800 years. The constellation Sagittarius, commonly called the archer, is illustrated in the Lost Book as having its arrow pointing directly to the middle of our galaxy. The arrow points to the center of the Milky Way. This galactic alignment, supposedly, is what will bring about the apocalypse that will end our world as we know it.

It's only been in the past five years that a black hole was found to be in the center of our Milky Way galaxy. We are not sure what will actually happen when this alignment happens or what influence this black hole will have on planet Earth. Some say it may make the Earth change its polar axis and swing into a different orbit thus causing horrendous events to occur. It has been proven that the Earth has experienced changes in the polar axis before over five thousand years ago in one of the extinctions we have documented. Others say this alignment will throw us into a different orbit and by doing that will head us into a collision course with a star or planet causing our immediate extinction and total destruction.

The "Lost Book of Nostradamus" contains eighty separate watercolor illustrations filled with all kinds of symbolism. This book is a countdown to the end of days. The illustration that is most disturbing shows the Book of Knowledge open with no writing on its pages. Does this illustrate that we may write a new life's record in the Book of Knowledge or does this show us that there is no life or knowledge to enter because of the impending apocalypse?

During the final days of Michel Nostradamus, it is thought the arthritis in his hands prevented him from continuing to write. It was at that time he hired a secretary to record his visions. This just adds to the mystery of who did the more primitive illustrations in the "Lost Book" for him. Was it his young son, Cesar, who later became an accomplished artist or was it his secretary making a crude attempt to depict on paper the visions made by Michel? If Cesar was only eleven when his father died, wouldn't the drawings tend to be rather primitive until his talents were developed?

Nostradamus was only one of the many prophets recognized in our past. He preferred to be known as a prophet and not a seer. The difference being that he felt a seer would be more like a palm reader rather than someone who made predictions without having any previous knowledge of the event. I guess he has earned the privilege of being called a prophet by many.

Interpreters of the quatrains credit Michel with some

of the most astounding predictions.

It is said he predicted the rule of the Emperor Napoleon, and even the World Wars, and Hitler. The events of September 11, 2001 are described in much detail as are many other important historic events. Of course, the one most disturbing at present and most timely is the war in the Middle East and the eventual destruction of our world.

The mystery continues and scholars are frantically studying these Medieval writings of the times to discover everything they can extract from his books and illustrations. Will they be in time to change something or is the die cast for us? Would we be able to reverse our plight? Are we doomed for extinction from the heavens?

Chapter Two

THE MAYANS AND THEIR CALENDAR

Where did the Mayans go? Where did they learn all their advanced knowledge of the stars? Why were they so obsessed with the stars? Why do some of their carvings include men looking as though they are in a space suit in a space capsule or rocket? How could they be so accurate with their calendar? How did and why did their buildings and temples resemble the pyramids?

The Mayans have remained another of the greatest mysteries of all times. How could such an advanced civilization just have one day disappeared? Did someone come and get them? If they relocated, then where did they go? Why did they leave? Some say they poisoned and polluted their natural supplies until there was nothing left for them to use, so they left.

The majority of the Mayan population did totally disappear, because there are people who claim to be direct descendants who are alive today, they didn't all disappear. Descendants still celebrate some of the ceremonies and rituals of their ancestors.

In one of the pyramids at Palenque there is a

sarcophagus picturing a man leaning back in a declined position in a seat. There appears to be some sort of equipment he is sitting inside. His foot is resting on what looks to be a control pedal. There appears to be a tube for oxygen and buttons on his garments like a spacesuit might have. From behind this mechanical looking machine are flames shooting out from. You would swear he is in a space capsule if you viewed it in person. How could image this be carved into the sarcophagus unless it was something they had witnessed? Could this have been in the imagination of the sculptor? I don't think so.

It is said the calendar of the Mayans, called the Mayan or Meso-American Long Count Calendar, is one of the world's most advanced calendars with accurate calculations of all sorts of celestial events. They believed time was divided up into particular blocks of time, cycles or ages, and related them to events that were to occur. There is much said about the ending of time, according to their calendar calculated to be on December 21, 2012. What did they see in their calculations? Why would their obsession with the stars and their gods lead them to this conclusion? They claim the last cycle was 5200 years ago.

There is much controversy as to whether the

The Meso-American Long Count Calendar

calendar is denoting the end of the world or the end of their last cycle and the beginning of the new cycle. Whatever the conclusion is we will soon find out whether it's an end or a beginning.

In a village of Teotihuacan, the pyramids and surrounding buildings are arranged directly in the order of our solar system. Many astrological events are documented and accredited to the Mayans. The pyramid built there is exactly the same size as the Great Pyramid in Giza. In fact, the alignment of the three buildings there are exactly the same. The sides are aligned with the points on a compass. How can that happen when they were built on two totally different continents? Did they possibly have the same designer?

In a city called Copan there was actually a written language. Stories were told by writing on large stones and were often signed by the storyteller or artist. Upon looking closer at these stones, one thing that strikes you as being very odd is that some of the figures depicted look like Asians, specifically the Chinese. How could they have known what they even looked like? There is much said about whether there were visitors from one side of the world that were brought to the other. How else would they know what an Asian looked like?

Also, in Copan, the steps leading to the top had carvings on every single step. This was a fairly large city with approximately 20,000 citizens so there were plenty of people to do these carvings. What do the carvings tell us? They are being studied as we speak.

The people of Copan completely vanished. Maybe these carvings are a story of where they went and how.

The Mayan religion had some very cruel and bloody rituals with blood sacrificing and bloodletting to the gods. It is said some of the humans were sacrificed having their hearts ripped out while they were still alive. Blood played an important part of the majority of their rituals.

Chapter Three

THE HOPI INDIANS OF ARIZONA

The Hopi Indian Tribes of Arizona are much like the Mayans in that they share the predictions of an impending apocalypse. The meaning of Hopi is "good, wise or peaceful people", but their forecast for the future is anything but that.

The Hopi Indians believe the earth has witnessed four purifications. They believe we are getting close to our fifth purification. Each purification has ended with the destruction of earth from a celestial event. They believe there are nine signs that must happen as a precursor to this fifth purification.

These are the nine signs of the Hopi. This is quoted from an account told by a Hopi Indian, named White Feather, to a minister named David Young back in 1958.

"This is the First Sign: We are told of the coming of the white-skinned men, like Pahana, but not living like Pahana men who took the land that was not theirs. And men who struck their enemies with thunder.

"This is the Second Sign: Our lands will see the coming of spinning wheels filled with voices. In his youth, my father saw this prophecy come true with his eyes -- the white men bringing their families in wagons across the prairies."

"This is the Third Sign: A strange beast like a buffalo but with great long horns, will overrun the land in large numbers. These White Feather saw with his eyes -- the coming of the white men's cattle."

"This is the Fourth Sign: The land will be crossed by snakes of iron."

"This is the Fifth Sign: The land shall be criss-crossed by a giant spider's web."

"This is the Sixth sign: The land shall be criss-crossed with rivers of stone that make pictures in the sun."

"This is the Seventh Sign: You will hear of the sea turning black, and many living things dying because of it."

"This is the Eight Sign: You will see many youth, who wear their hair long like my people, come and join the tribal nations, to learn their ways and wisdom.

"And this is the Ninth and Last Sign: You will hear of a dwelling-place in the heavens, above the earth, that shall fall with a great crash. It will appear as a blue star. Very soon after this, the ceremonies of my

people will cease."

According to the Hopi, seven of these signs have already taken place. The seventh sign is probably referencing the oil spills throughout the world.

In their legends they speak of the Ant People. At the time of each purification, if you have been chosen, you enter the center of the heaven and the Ant People will take care of you. These Ant People look very similar to how the "grays" are described according to witnesses of some extraterrestrial encounter.

There is a very sacred area, not open to the public, that has many drawings depicting the history and tales of the Hopi. On one of these, called the Prophecy Rock, there are many drawings of strange creatures that look as though they are wearing space suits. Others, the Ant People as they call them, look like what we would call "grays". There is even what looks to be a flying saucer. How do you explain that?

The end will be preceded by the god, the Blue Star Katchina, removing his mask and dancing. When the Blue Star appears, signaling the end, there will be many catastrophes like major earthquakes and massive flooding. Just before that event, the chosen will be taken into the center of heaven once again to be saved.

The Blue Star was just recently discovered by astronomers. Scientists actually found two blue stars together, but as they came near the center of the galaxy, where there is a black hole, they joined into one. The blue color represents a dying star. We are in no imminent danger from this blue star according to most scientists. Scientists have been wrong before.

One fact that was mentioned about the blue color was that some comets can look blue, too. They seemed to be more concerned that we are close to the asteroid belt than being worried about the blue star.

By doing core samples of different areas throughout South America, evidence has been found of what may have been a polar shift 5200 years ago. There would have been such devastation. There would have been floods and earthquakes and finally the world would have been frozen over.

The Dogan tribe in Africa has a legend about a blue star, too. As did the Zuni Indians.

Chapter Four

OTHER ORACLES AND PROPHETS

All ages have had their seers and sages. We have always been obsessed with wanting to know what's going to happen. Going back as far as the Celtics and Druids there have been these oracles who reportedly had direct communication with the Gods. They were advisors to the elite and rulers of many civilizations and their word was not to be questioned. Oracles were said to have prophetic connections with the gods.

The Greeks had many oracles, but three were the most recognized. One of the oracles was feminine, Pythia. She was said to have eaten laurel leaves before her ordeal of going into a trance and uttering words of wisdom while she was in it. Her visions were, also, accompanied by breathing the gases from fumes from a volcanic crack in the ground. She would shout out words that sometimes made no sense, but that would then be interpreted by a council.

Two other male Greek oracles were Dodona and Trophonius. They, too, advised the ancients. Before going into battle or any important event the rulers and advisers would be preceded by having a consultation with an oracle. Their words were golden.

China had its I Ching. The seers used three special coins that had certain markings on them and they would be tossed six times. The results of the tosses would be recorded. Then a book would be consulted and the reading would proceed. It worked by using sixty four different combinations of the way the coins could possibly fall. It's rather disturbing that their calendar shows their time line ended on December 21, 2012. Modern day astrology still offers a reading using I Ching.

Sybil, an oracle in the times of the Romans, lived in a cave and inhaled volcanic fumes to enter a trance, then she would have her consultation with the gods. Her predictions, written on oak leaves, were centered around time being divided into a blocks of eight hundred years. She said their would be nine of these periods , but then the tenth period, calculated as beginning in 2000 AD, would be the last. The end of times would result from an apocalypse. Many of her predictions were revered by very prominent citizens including Michelangelo. The coming of the birth of Christ was, also, one of her forecasts.

Hinduism had its oracles, too. The Akashwani was their contact with the gods. And just as with all the oracles, the words were sacred.

Even in the present day, the Dalai Lama of Tibet consults the wisdom of an oracle.

The Norse had Asgard to speak with the gods and advise them. They predicted the world would have a catastrophic event that ended with the world being covered with ice.

Hawaiians, having their oracles to consult with, have celebrations and rituals honoring these oracles even now.

Sir Isaac Newton spent fifty years of is life looking to break the code in the Bible.

There is a present day mathematician, Bruce Bueno de Mesquita, who has formulated a code that has been extremely accurate. Because of his accuracy, he has done work for our CIA and numerous other agencies. Does he know when the end is to come?

20

Chapter Five

THE SUMERIANS AND EGYPTIANS

As far back as 6000 BC, is found the oldest culture known to us, the Sumerians. Not only were they credited with the first written language , but their system of mathematics, which was based on units of ten, became the foundation for our common day mathematics.

Their cities in the Southern Mesopotamian region, had a sewage system, cobblestone roads and were laid out like our cities are today. They were very knowledgeable about agriculture and had irrigation systems. They, also, had an acute knowledge of astronomy.

The story of the Sumerian civilization was recorded on tablets. There were several thousand tablets found and the code was broken so they could be interpreted. A man named Zecharias Sitchen wrote a series of books about what he uncovered.

According to Sitchen, the Sumerians recorded a visitation from the "Anunnaki", which means those who from heaven come to Earth. These Anunnaki, also called Nephilim, were giants from the Planet Niburi.

The planet passes by the Earth every 3500 years. It might be noted at this point that it takes 3600 years for a complete revolution of our galaxy.

These Anunnaki came to Earth to mine for minerals and ores, mostly gold, according to Sitchen. Upon their arrival and needing slaves to do this mining, but only having the Neanderthal as workers, they created man from the Neanderthals. They, also, created other creatures to help them. Then they destroyed the Neanderthals. The creation of these first humans was for the sole purpose of them to be used as slaves and as servants. After the mining was completed, they left for their home planet. Some of the Anunnaki stayed to oversee the slaves and settled here.

Several years ago, in Greece, archeologists uncovered the remains of a giant species that looked humanoid. These humanoids were seven to nine feet tall. Among some of the artifacts found in the dig was a writing with the name Goliath on it. Could this have been the Goliath David slew in the Bible?

In Africa, there have been mines found to be over one hundred thousand years old. Who mined them?

In the book of Genesis 6:4 is written, "There were giants on the Earth in those days." The Anakim, were a race of giants mentioned in the Bible, too. In fact, the descendants of Cain were have said to have entered into sexual relationships with the Sethites and they had children who were said to be gigantic.

In the Lost Books of the Gospel are many stories of the intermingling of giants an humans. The Lost Books were Books not accepted to be included in the Bible. It was thought the books contained too much detail about things not accepted by the religious sects of the times.

We can thank the Sumerians for being able to decode the hieroglyphics of the Egyptians.

Probably one of the most discussed and theorized about topics of our ancient past is still the Egyptians and how the pyramids were built? The stones are immense and weigh tons. The manpower to relocate them and place them alone would have been mind boggling. Even with our modern technologies, with all the equipment we have available at our disposal, it would be extremely difficult to build these ancient structures today. Did they have help? Did more than just the Sumerians have help from the gods

What makes us believe that we are so unique as to be the only civilized and developed planet in this huge universe of ours? Or in all the other universes in space? Why couldn't we consider tht just maybe Earth could have been visited before and previously mined? Could we be the only planet that advanced to this point? Wouldn't that be rather arrogant of us to believe we're alone? Could they have been here before to help with our advancement? Are our technologies from visitors from beyond, only trying to help us move forward more quickly?

The largest of the pyramids in Giza stands 471 feet high. We believe it was constructed around 2560 BC. Each of the four sides of the pyramid are aligned with the four points on the compass. There is not one marking on the Great Pyramid of Giza of any kind. So then why was it built? An inner burial chamber has never been found.

How were the pyramids built? How were those gigantic stones quarried, stacked, and carried to the top? How were the hieroglyphics able to be written or carved into the interior walls of the other pyramids when there was no light and not enough oxygen inside to even keep a candle lit?

What was found was four passageways that were perfectly aligned to point to specific constellations. Why were there secret passageways that go nowhere?

What was their purpose? Why do the pyramids built to line up in an astronomical alignment with the certain stars, planets, and constellations? What are the strange creatures pictured in the hieroglyphics on the walls and hallways of the pyramids? What are the strange devices pictured there?

   A recent documentary I watched was showing the tunnels and passageways that were hidden inside the interior of the Great Pyramid. They sent a small robotic device equipped with a camera attachment and placed it inside this space. Other tunnels and other doorways were discovered. These tunnels would dead end into doorways that ended up with another door to open beyond that. The Director of Antiquities in Egypt is just now allowing the drilling of a hole in these doorways to try to discover exactly what is their purpose and where they go. What will they find?

   There have been many new discoveries found in the area surrounding the pyramids, too. The riches uncovered, the stories illustrated on the walls, are beyond our comprehension. A full sized boat was uncovered in one pyramid and it was in wonderful shape after all these centuries. The vessel was for easy passage for the pharaoh from this world to the world in the afterlife with the gods. Egyptians, as we all know, were very careful to include all the items they felt they would need to function in that afterlife. The bodies were carefully prepared and the vessels containing their organs were buried with them. The gold and precious jewels and carvings placed in with them illustrated the wealth they must have had during their lives on earth. There was a scroll buried with them to be read that would bring them to life in the other world.

In the Book of the Dead" is a prediction that when the Planet Venus loops above Orion there will be a great catastrophe. This is to happen in 2012. It goes on to say that " after the destruction, the old lion turned around." Did they mean there would be polar shift?

What an incredibly intelligent and innovative civilization they must have had. They left, for us to see, messages and proof of how they lived in their hieroglyphs. Their messages were left on the walls and on tablets for us to interpret and read for a reason. In these drawings we see creatures and figures that don't look like any kind of creatures we would be able to recognize today. Were these something they actually witnessed? If so, where did they come from and where did they go? Were they genetic experiments?

Think about this for a minute. It has been proven that deep within the interior of the hallways in the pyramids, the oxygen level is so poor you can't even light a match. The theory that copper mirrors were set up to help transfer light from the surface down these passageways just doesn't work but only for a short distance. As you get further down, the light can't keep its strength. How did they stay down there to make these drawings? How did they breathe down there? Did they have some advanced technology shared with them from somewhere else? Why is there no evidence of soot from torches if that's what they used?

The Gods that were worshipped were said to have come from the heavens. Many drawings show the sun and the planets and figures seemingly coming down from the sky. Why were these gods pictured as looking different from what the normal Egyptian looked like? Why did they tell the story that the gods gave them the knowledge for a lot of their accomplishments? Why did their gods give them this knowledge?

Could we be talking about knowledge given from an advanced alien visitation? Was that what gave them the ability to build the pyramids and cities and statuary to last through the ages? One has to give some thought to whether this should be a reasonable consideration.

Much has been said about the alignment of the

three pyramids of Giza to certain constellations in the heavens. Not only do they line up, but so do other earlier pyramids and temples built throughout Egypt. Was this meant to be another message to us? Are they pointing to where we should be looking to find our visitors?

How is it that there are literally thousands and thousands of huge statuaries in the cities? How is it possible they could have sculpted these figures from stones weighing many tons then have transported them to their place in the temples? Try taking a block of granite weighing ten tons and chiseling it to a block with perfect measurements with the tools available in that time. How long would it take you? Even the limestone would require a lot of time to carve out. How could they have possibly physically done it?

Did they have special tools or some type of advanced technology brought from somewhere else for them to use, but not left behind for others to find? Why haven't archeologists found millions of carving tools? The tools found are very primitive for what you would have expected them to use to shape the blocks and statues.

The top, capstone, of the Great Pyramid was said to

have been covered with gold, which has long since disappeared. Since gold is a great conductor of electricity, why would it be placed on top of the pyramid unless it was meant to be a beacon light of some kind or used to transfer a power of some type? Maybe this was supposed to be like a lighthouse for guiding others to this particular spot on earth. Does that make sense to you? What was the purpose for this? The pyramids exterior was covered with highly polished limestone and said to be one of the few things on earth that could have been seen from outer space. I'm not sure that's really true though. But if you could then why would that be so unless it was intended to do just that? Intended to be a guide.

There have been some theorists dedicated to proving that the pyramids were some sort of huge machines used to produced an energy ray like a microwave. The thinking is that if they had this microwave signal going into space, it could be redirected and sent to another place on earth and used as an energy source. Isn't that very thing happening today? The earth has a power grid. Was this supposed to tap into that grid somehow?

Another theory is that the pyramid was a huge converter taking minerals exactly as an alchemist would have, to actually make gold. Why gold? Not only is gold a good conductor of electricity, but it doesn't corrode, and it is a perfect reflector of infrared heat. A recent discovery of channels and canals under the Sphinx is thought to be a repository of some kind. Gold maybe? What if they were only here to mine gold? What if gold was needed where they came from? Were the Egyptians only meant to be slaves to them to build this gigantic converter?

Yet another recent theory is that the Great Pyramid was left for us to be used to destroy the asteroid that will cause our destruction. The idea being that it can function like a laser beam if you know how to use it.

And what of the strange creatures pictured? What if the weird creatures pictured in the hieroglyphics really existed? We see there in their drawings combinations of creatures like the half horse with the other half man, the heads of birds combined with man, and many other combinations of different animals. Were they experimenting with these combinations to find the most productive or strongest animals? Were they helping the Egyptians drag these huge tones into place?

What if these creatures were experiments done by genetic engineering? What if these creatures were the gods themselves? Why have we never found the remains of any of them? Were they once living or visiting with the Egyptians and when they were done with whatever they were here to do, did they leave with the gods? If they were just dreamt up by someone's imagination then why are they everywhere? Could the entire society have dreamt up the same figures? What if they were like pets of the extraterrestrials? Is this so far fetched?

Actually, we may have some of the remains found in huge stone burial box found in Egypt.

Today we implant the seeds of a male donor into women wanting to have a baby and not able to conceive on their own. Could this have been possible centuries ago?

We have genetically created sheep, mules, lizards, from altering the DNA by cloning. If we had gone up to an American Indian, for instance, and told them we could duplicate their favorite horse, would they have believed us? This would have sounded as ridiculous to them as the theory that creatures shown on the walls of ancient Egypt were real, too.

The Egyptians truly believed they came from the stars. One of their Pharaohs, Akhenaten, in 1352 BC, was the pharaoh said to come directly from the stars to live among the people. His influence of religion changed their belief in many gods to having only one god to worship.

Akhenaten's body was strangely shape with an elongated head, big belly, large hips and square shoulders. He ruled for less than twenty years. His son is one of the most famous and most recognizable of the pharaohs, King Tut, Tutankhamun. His mother was Nefertiti. And don't forget that King Tut's tomb was filled with treasures of gold beyond the imagination. Maybe he was going to take them with him alright.

The Egyptians were so advanced with their civilization, isn't there room for some question as to how they got that way so quickly? Why didn't they leave us better records? Maybe they have.

You could fill an entire book with all the questions that still go unanswered about the Egyptian civilization. When you think you have the answers to one thing, you uncover yet another thing that in turn makes you ask more questions. Every year something else is found that totally changes what we previously knew. History is constantly getting rewritten because of what we uncover. Will we ever know the answers? Maybe sooner than you think.

Chapter Six

THE SKULLS OF THE ANCIENTS

As I have mentioned before, the ancient Egyptians had a pharaoh they believed came from the heavens and the stars. Sirius is the star they point to as to being the birthplace of the Pharaoh Akhenaten. This King from the stars was depicted as having an elongated head and strange body. He is pictured wearing head-dresses that were pointed to help to accommodate this misshapen head. As did his wife, Nefertiti.

It is told that after only ruling for seventeen years, he disappeared and left this kingdom. The statues made of him were destroyed enough to conceal what he really looked like. It is only because small statues of him survived that we are aware of this deformity. After his rule, the worship of many gods was again implemented.

He, also, left a son behind who was probably the most well recognized pharaoh of all ages. His name was King Tutankhamun. After the remains of King Akhenaten were found, it was discovered that he indeed had an elongated head. What was startling is that King Tut had this same elongated skull. So did

Akhennaten's other children.

This elongated skull is present in cultures all over the world. The Dogan, an African tribe in Mali, actually wrap the heads of their babies to make it conform to look elongated. Was this to keep the shape going on for eternity? Where did they see this shape? Was it from an earlier visitation? After all, they claim to have had their gods come from Sirius, the same star as the Egyptians. How could they have known about a star not found until recently by astronomers with advanced equipment?

The masks of the Dogan tell the tale of these visitors from the stars. These are ancient masks used in their ceremony. Every year there have a ritual ceremony called the Festival of the Visitation. This tribe is literally in the middle of a desolate place. How would they know about any of this except through the stories carried through the ages? They had no written language.

In China, around 3000 BC, their god was said to have ascended from the heavens from a Great Star. This god is said to have arrived on earth on a flying dragon. During his rule, many great discoveries were made like the invention of the compass. He was, also, said to have left for the heavens on this dragon. There are many strange looking creatures in their paintings. Some are birdlike with elongated looking skulls.

There are tribes living throughout South America who have been shown to have their heads misshapen. If you look closely at the Mayans, you will notice their heads are not normal size and their noses are pictured as looking larger than what we would consider as being normal.

Skulls are wound into many Mayan legends. For instance, the Legend of the Thirteen Skulls is said by the Mayans as to have great significance with what happens to our world. It is said that when the Thirteen Skulls are reunited a great knowledge will be unveiled. They tie this legend in with the end of the Mayan calendar. They are supposed to be the depository of great wisdom. They have never been accurately dated as to the actual age. These crystal skulls are spread all over the world. What must they mean? How do we receive the knowledge they are supposed to have? When the thirteen skulls are united, then some great power is to be released. What could that power be and how will it effect us?

The Incans of Peru, also, had elongated heads. Their legends and stories refer to the Sky People or Star People. One of their stories tells of the "tracks" left behind by the Star People. Were these the Lines of Nasca they were talking about? The geo-glyphs had picture of strange looking creatures wearing what looks like space suits.

These misshapen skulls could very well be like what the "grays" that are reported as being seen in sightings resemble.

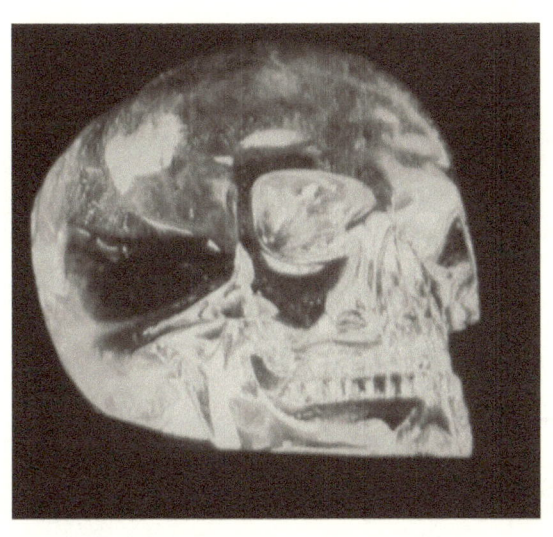

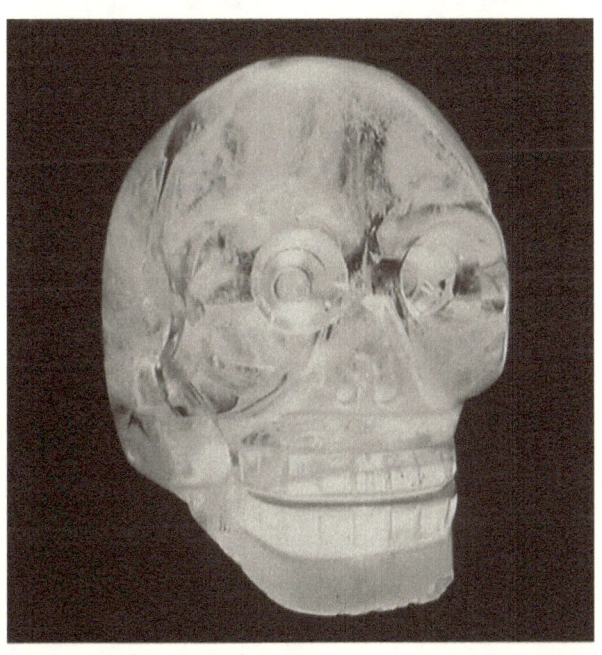

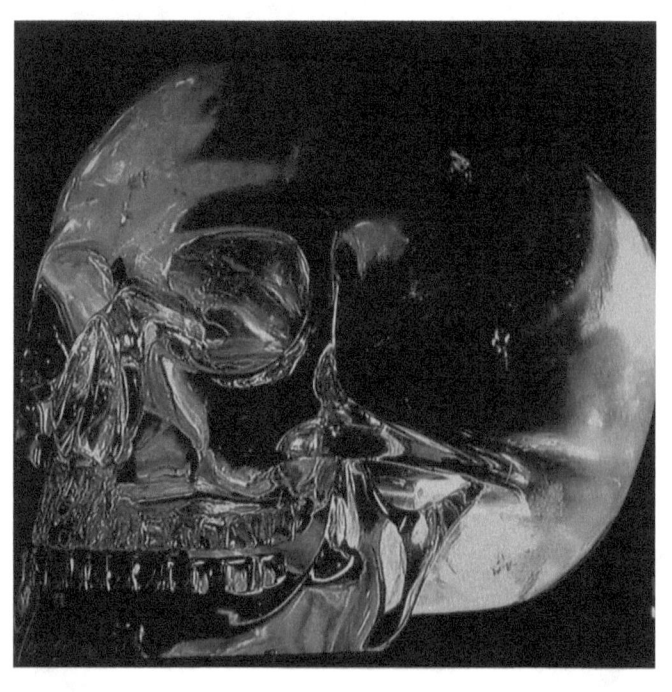

Chapter Seven

STONE MONUMENTS OF THE WORLD

Why was Stonehenge erected and who built it and why was it constructed? Why were the stones put there? What purposes did the stones serve? Why do we find similar structures throughout the world?

I think today the popular opinion is that the Druids, Greeks, Phoenicians or Atlanteans erected England's Stonehenge. More recent information indicates that it was not built all at once, but in several stages form somewhere between 2800 to 1800 BC.

How were these huge stones stacked on top of each other? How were they obviously moved around to be arranged in a circle? Why a circle? Why circle after circle?

If you have ever actually visited Stonehenge, you will find yourself out in the middle of a field in the middle of nowhere. I'm not sure what I expected to find, but its placement didn't seem to make sense to me at all.

As you go on the guided tour, the guide points out just how very large these stones are. Maybe I could envision how the stones could be moved there, I don't know, but how on earth did they stack the larger ones on top without some type of machinery? These are huge heavy stones. Even the smaller blue stones, as they are called, would be difficult to move and place. What did they move them with, especially long distances? Where is the evidence of a crane or tripod?

Some people say they can feel some sort of energy radiating from them. Maybe they can. I didn't notice anything like that. That's not to say it doesn't happen. You really can't see energy fields.

I walked around the stones, taking many pictures, so I could look at them later. It is quite a puzzle as to what they were for in the first place. There has been lots of information coming to the forefront that it was tied to a religious event of some kind. The tour guide will tell you that there may have been sacrifices made there and then they point out where the altar might have been.

Standing there looking at them made me think of it as being a sturdy base to put something very heavy on top. This would have held something very substantial.

It would have been real overkill to have put a roof

over it so I doubted that. I kept wondering if it could have been a platform for something. Could it have been a landing pad of some kind? Why would it have to be built out of such huge stones if it were used for religious ceremonies?

Every year there is a celebration and a ceremony celebrating the summer or winter solstice alignment seen through the stones. It is said that it has an alignment with eclipses, too, making it an astronomical phenomenon. Was that its sole purpose? Can you imagine that a civilization so primitive would even bother looking up at the stars? I could maybe justify them wanting to understand the movement of the moon or sun, but that's about it. Why such a fascination for the stars?

Not far from Stonehenge is another area where you will find huge monolithic stones. It is in the town of Avebury. I went to see these, too. These were much more recently placed by the farmers in that area. There history is at least known. I never felt any energy fields there either. You are free to walk among these stones so I have lots of pictures of them, too.

During my visit to Avebury, the local people spoke of the crop circles found in fields nearby. My time was limited so I didn't pursue this any further. The locals claim that the whole area had an energy field around it and that's why there were crop circles there. They had books filled with pictures showing these crop circles and others in England. A lot of these were proven to be hoaxes, but some are still being investigated.

Europe has many Stonehenge-like structures and they garner the same amount of mystery surrounding them that Stonehenge does.

There is another "Stonehenge" in Armenia, said to be 7500 years old, where the stones have holes bored in certain spots. Why do these holes in the stones seem have an direct astrological alignment with the constellation Cygnus? Why that particular constellation? The constellation is sometimes referred to as the Northern Cross, but is commonly called the Swan as derived from Greek mythology.

America has its own Stonehenge in Salem, New Hampshire. The actual dating of this has had a lot of controversy concerning it. There are many theories as to what it was used for, also. It's still be heavily investigated and studied.

Monolithic stones are found all over the world. Some of these feature stones so closely placed that you can't place a piece of paper between them. No mortar was used either. How is this possible without modern equipment?

There is a stone called the Thunder Stone located in St. Petersburg, Russia that would be trying for our modern equipment to move and set in place.

The pharaoh, Ramses II, had a statute of himself made and transported from 170 miles away by ship. This statue weighed over a thousand tons. How did they manage to first move it to a ship then unload it somewhere else?

It is thought that the ancient Greeks and Romans had the use of treadmills and cranes of some type. If this equipment was available in the times of the ancient Egyptians, why were they not pictured anywhere? Why would we never have found one of these cranes or treadmills? True, the timbers would have rotted away, but the illustrations wouldn't have disappeared.

There are the Carnac Stones in Brittany, France. These are probably the largest collection of stones in the world. Said to have been erected by a Pre-Celtic people, as far back as 4500 BC, and numbering over three thousand, they are arranged in long rows. The measurements from end to end and side to side are perfect. Local legend is that a magical spell was cast upon soldiers and they were turned to stone. Can you imagine how these megaliths were placed?

   Here, in the Southern United States, is an arrangement of five fieldstone slabs, standing twenty feet high, placed in a field in Elbert County, Georgia. There is an astronomical alignment with the solstices and eclipses. Built in 1979, these stones have Ten Guidelines left for humanity and inscribed in twelve different languages. This message is supposed to be the guideline for us to follow for right after the Apocalypse. One of the Ten Guidelines is to hold the population of the world to 500,000,000. This is so that we will be in balance with nature according to that guideline. What happens to the remainder of the people? It speaks of a new world order in another line. Does that sound familiar? Is this because we are all going to be under the same "God" finally?

   Why are so many messages left for us to see made of stone? Stone is the forever medium. It should last virtually forever. The pyramids all over the world, the statues, temples, cities, all built of stone. Stone left throughout eternity for us to find and interpret their meaning.

Chapter Eight

ANOMALIES LEFT BEHIND

Why are there strange statues on Easter Island? Why are they looking upward to the skies with their backs toward the open waters? What are they looking for or better yet, what are they waiting for? How did they get there and how were they carved? Why do they look like people in the carvings of the Mayans? What happened to this whole civilization? Where are the trees on Easter Island? It was once a thriving island with lush plantings.

Can you explain what the lines in the sands in Nasca, Peru mean? Why would the Incans draw these lines where you couldn't see them from the ground? Why do some look as though they were meant to be a runway? What do they mean? Who are they meant for? One of the mountain tops had been flattened off, but there's no evidence of where the dirt went. Of all the places in South America, this area holds examples of most of the elements and minerals that comprise our world. If there was mining going on then an airstrip might be needed for the ships to transport these ores.

How could a map of the world been drawn with extreme accuracy in the early 1500's when the only way you could do this was from an aerial view? How could the Antarctic region be pictured in such detail from the ground view? More than that, how could it have been pictured as representing what Antarctica looked like when it had no ice? That would have been millions of years ago.

How is it that on the Island of Galapagos there are some of the strangest creatures on Earth? How is it that there are animals that are found only in certain particular areas of the world seemingly to have been separated for some unknown purpose? Or maybe for controlled breeding? Maybe they were the rejects.

What happened to the dinosaurs? One of the older theories about an Ice Age has since gone by the wayside. Some theories say bacteria and viruses could have been the culprit. What if they were found to be too primitive to be of any use? Maybe they were deliberately zapped away.

A super volcanic eruption is yet another theory being studied. This, too, has some problems because of where the fossilized remains are being discovered. It doesn't stand to reason how they are found together.

Where's the missing link between apes and the humanoids? There seems to be a gap in time for what should have been the evolution from apes. How did we get from being cavemen with very primitive tools to discovering fire and the wheel? Did we have a little genetic engineering to help us develop our brains more quickly? Did we have a visitor who showed us what to do with fire? Can you imagine how quickly it seems we went from cave dwelling to building pyramids?

The Sumerians, of 3800 BC and back as far as 6000 BC, had one of the most advanced civilizations of all times. They even had a flushing toilet. Their written language was one of the first and their mathematics was based on units of ten as we have today.

There was a discovery of literally thousands of tablets found that had stories of the Anunnaki, which translates to "those who from the heaven come to earth. This is a direct reference to beings who came from the stars. An actual eyewitness account. Where did the Anunnaki come from? They claim they are coming back according to the Sumerians.

Why are there stories in the Bible that tell of powers unbelievable for those times? Like powers capable of total destruction of a city or of armies. Is the Ark of the Covenant a vessel that held something that possessed the power to destroy full armies or cities? It is stated in the Bible that it had powers unimaginable. The mere accidental touch from a carrier caused horrible burns and death to others. Was there some radioactive energy associated with it? Was it a radiation sickness of some kind? And what about being able to speak to God with it?

How were the seas parted by Moses? Was the staff really some kind of energy force given to him by God? What about the burning bush? A bush in flames that is not destroyed by the fire?

How do you explain manna from "heaven?" Did they have supplies dropped from heaven just like we do our modern day air drops? It had to come from somewhere. Why not a spaceship?

What about the story of Sodom and Gomorrah and Lot's wife being turned to a pillar of salt? The actual site has recently been uncovered or at least that's what the thinking is at present. Was there a nuclear explosion of some kind and is that why God told them not to look back? If you look directly into a nuclear blast, the brightness is supposed to be blinding. Is that what happened? Maybe the "Raiders of the Lost Ark" didn't have it so wrong.

And what of the Immaculate Conception? Was Mary implanted with the seed of Jesus just like we do artificial insemination today? Is it so unbelievable? If she was, this would in no way diminish her role in the birth of Christ and the Christian religion. Maybe the bright star that led the wise men wasn't a star at all. Maybe it was an extraterrestrial vehicle sent from God.

How is it that the Bible isn't the only manuscript to speak of untold powers, gods from the heavens and winged creatures? How is it that the stories are gradually being proven to be true slowly but surely? Does the discovery of the extra books of the Bible add insight to any of this? How about the Dead Sea Scrolls? What are we not being told because of the fear it would change the way we believe in our different religions and cause chaos?

The newly discovered scrolls and forbidden extra books of the Bible contain all sorts of references to the gods from heaven. In our day of computers and advanced technology, why haven't we been able to view the contents of all written documents that have been found? If the Pope in 2008 made the statement that if other extraterrestrials are found, that we should continue in our God. He said it should not contradict any belief in God. What does he know that we don't?

There could be all kinds of ancient documents held by the Vatican in its Treasury. Only the Vatican would decide what to do about them.

If you examine many of the paintings and art of the ancients and medieval times, you will see things in the sky in the backgrounds on some of these works of art. How is it many of the older paintings, woodcuts, and drawings from our past artists reveal some strange mechanical looking devices in the skies? Is it from something they actually witnessed? Could they have been imagined? All of them?

Have you ever looked at the drawings for some of the inventions of Leonardo Da Vinci? Could he have been helped a little to make these sketches or was he really so brilliant? Could he have witnessed something in the skies to prompt the drawings of the flying machines?

How many inventions have been made that were of such importance as to have a major effect on our society? Look what has happened in the past twenty five years alone. Do you remember how large the first computer was? And how awkward and heavy the cordless phone was? How about how thin out televisions are now? Look at everything we have in our homes. The microchips made now can hold a vast amount of data.

We have made enormous advancements in every field in the last century. Have we had a little assistance along the way? How did we go from the caves as hunters and gatherers to where we are today in just a blink in time?

Easter Island

Chapter Nine

THE BOOK OF REVELATIONS

I seriously doubt that any other book in the Bible has been examined as closely as the Book of Revelations. For centuries man has tried to interpret and understand what is being said and what it means.

The word "revelation" means the revealing of something previously hidden or secret. It can, also, mean a surprisingly good or valuable experience. What is going to be revealed to us? With a possible apocalypse at hand, is the promise of a Utopian world and the promise of the return of Jesus after the destruction the good experience we can hope to have?

Every month there are discoveries of artifacts from older civilizations we weren't aware of or some astronomical anomaly that changes our previous notions. We sometimes are able to decipher and understand old documents due to the advent of the computer.

The Apostle John is who is credited with writing the Book of Revelations. John was banished to the island of Patmos by the Romans who were persecuting every believer of Christ.

John was visited by an angel who took him to God. God showed John a scroll that had seven seals and God told John of the horrors that were to take place at the End of Times. The Four Horsemen of the Apocalypse appeared in a vision. The first horse was White and represented conquest. The second horse was Red and represented War. The third horse was Black and it represented the economic collapse of the world, hunger and famine. The fourth was Pale and represented Death.

Some believe there will be a Great Rapture where the saved will go up into Heaven. These will be the chosen.

The Lamb, Jesus, appeared with the scroll and four of the seals were broken releasing the horrors on Earth. Jesus banishes the Whore of Babylon who represents Rome. The seventh seal brought seven angels. It is when these seven angels come that the end is very near.

The Beast with seven heads comes forward. The Beast is said to be Nero. The numbers of the Beast were 666 which is the numerical translation of the name for the Emperor Nero.

Some think there will be utter choas on the Earth.

Earthquakes, floods, famine, plagues, will overcome the Earth.

The Anti-Christ ill occupy the Temple of Solomon once again. He gathers his armies for the final showdown between Good and Evil.

The Battle of Armageddon is the place for this final battle to be fought.

Jesus descends from Heaven with the armies of God. He encircles Satan with fire and banishes him from the Earth forever.

The Earth will have Peace for eternity and become like a Paradise again with all the Earth renewed.

The Book of Revelations is the most difficult of all the Books of the Bible. Scholars disagree as to whether it is a literal account of what is to come or if it is meant to be symbolic.

It is not certain that John was really the author, but he is most commonly given credit for having written it.

It said the cave where John lived on Patmos was on top of a volcanic fault and gases escaped from there into the caves. Could that have influenced the visions he had much like the oracles of the past?

Still others believe the war will be an atomic one. With the Middle East in such turmoil today, this might not be so hard to imagine.

One of the last signs before the End of Times is the people of Israel occupying the land once again and the Temple will be rebuilt.

The Book of Revelations was originally not going to be included in the Bible.

My account of the translation of the Book is very much over simplified, but nobody knows how to translate it properly.

Chapter Ten

<u>BRIGHT LIGHTS IN THE SKIES</u>

From ancient times to the present, there have been reports of strange lights in the skies above. The Bible mentions strange machines and strange lights many times. And so does the Koran and the Torah. Time after time they refer to the angels descending from the heavens. Were these angels actually extraterrestrials? Could they have been beamed down on some sort of ray to earth? Weren't you taught as a child that the angels live in the clouds? Did they really have wings or would that be how the early civilizations have interpreted what they had seen?

In present day India, the people would not be surprised at the mention of flying machines. They had Vimanas described in the sand scripts that were flying machines. They have actual illustrations in great detail as to how they looked and how they worked.

The story of flying carpets are found almost everywhere in their drawings and paintings. These magic carpets could instantly transport the gods from one place to another. Could they have been referring to a spaceship?

Throughout history there are descriptions of beings from the heavens and of machines in the skies. Can we write these sightings off or should we realize that what they were seeing might be true and real accounts of extraterrestrials?

What are strange lights that can be seen in the skies on all continents? Are they all explained away? Was there an actual crash landing of an alien craft in Roswell in the 1940's? Are we being told the truth? Can we trust our government to tell us the truth? Are all the people who have seen these strange lights delusional or not very credible witnesses? What about the "missing" time reported by some?

Why is there such mystery surrounding the Bermuda Triangles around the earth? Where did the planes and ships disappear to and will they come back? How is it that with all our highly sophisticated equipment, we haven't found these missing planes and ships? Can it possibly be true that these missing planes and vessels enter a vortex of some kind like a wormhole and just disappear into them? Could they perhaps be disappearing into an area underwater that is a transporter of some kind where they are drawn down and are destroyed or teleported to somewhere else?

Do you ignore the accounts of some of our own astronauts of having seen "something unidentified" flying along with them in orbit or in their sights? Even our Presidents have told of eyewitness accounts of lights in the sky. How about all the reports by some commercial pilots having near misses with strange objects? Are they all nuts? They should command the respect of the public to be very credible witnesses.

Can all these happenings be explained away logically? Can we remove all of our prejudices and predispositions to think beyond what we are taught from childhood? Would you at least agree that there are an enormous amount of events and curiosities that have never been explained to your approval?

As a young child in the 1960's, I can tell you of many sightings I witnessed with other people. Let me give you some background information first.

My Father was a pilot and a veteran of the USAF and the RAF. He spent many hours observing the skies and maybe that's why he was so fascinated with astronomy. He could tell you what and where every constellation was. We always had a telescope set up and as a family we spent a lot of time looking at the stars.

Our home was in Ocala, Florida, centrally located in the state, and where I spent most of my childhood. We could stand in our front yard, if it was clear skies, and watch the rockets go by from launchings in then Cape Canaveral. All the neighbors would watch, too. This was the most exciting thing we had ever done or at least to been able to witness. The space program was so new and seemed so amazing and impossible.

Many times we would notice objects just hovering in the sky during those launchings. My Dad called the local airport several times to ask if there were planes in the area. He had a friend who was a controller and would tell him. He never claimed having planes or helicopters in the area. We thought since the objects were grouped together most of the time that they would turn out to be helicopters. Until we watched these extremely bright white lights shoot off with lightening speed and sometimes in all different directions.

That's when we started watching the heavens a little closer. After dinner Dad would go out and set up the telescope and folding chairs and we would watch the skies. We didn't see objects very often except around the period of time for the launchings. Then it was very active.

I remember one particular night when Dad came into my bedroom and woke me up. He took me to the window and pointed to five bright objects in the sky over the golf course. They hovered for a bit then they moved slowly forward until they split off from each other and disappeared. We watched them for some time before they disappeared out of sight.

You have to remember that I grew up with the threat of nuclear war with Russia, the Cuban missile crisis, and the Cold War. Bomb shelters were being secretly dug in the yards of friends and neighbors. We couldn't afford to build one but we sure were rehearsed as to what to do if war broke out. And, yes, we had the school drills where the alarm would sound and we would have to get under our desks. A lot of good that would have done.

We, also, had the Roswell incident happening years before. Movies like "The War of the Worlds", "The Day the Earth Stood Still" and many others. Flying saucers were thought of as being very real and menacing. We didn't know if there was going to be an invasion or not. All we did know is that we saw strange things in the skies that were unexplainable at that time. Do I believe what I saw was something not of this earth ? Absolutely. I'd bet my life on it.

Another thing I might share with you was that I had an Uncle Fritz who worked in the space program. Uncle Fritz was what we called an egghead back then. He truly was a genius. He, also, was on think tanks for big corporations. In fact, he lived in California, but they flew him to Connecticut and Texas all the time. He later invented some device for the Apollo, too. My point being that this man was very intelligent.

I'm telling you about Uncle Fritz only because of something that I overheard one time. My Grandparents lived in Lakeland, Florida and on one of the visits to see them, my Uncle Fritz came by to see them, too.

That night after dinner, I was sent off to bed so the adults could talk about adult stuff. This particular night, I had sneaked part way down the stairs and sat down on one of the steps to listen to what the men were talking about. I was maybe ten then.

My Grandfather was a Colonel in the Air Force. He and Dad and Uncle Fritz would get into discussions about crazy things or at least that's what I thought about it. I remember them talking about positive and negative universes. I didn't think that was so special because we had learned about positive and negative in school. No big deal there. I knew what a universe was.

The topic that caught my attention and that I remembered the most was about black holes and worm holes. I thought that was extremely funny and stupid to be talking about. Obviously, I remembered that so it made an impression on me. I had no idea what they were talking about regarding them, but I will never forget it. Their conversations would tend to get very heated and loud. Their discussions went on for hours and I would listen until I got too sleepy to stay up anymore.

Uncle Fritz, Dad and my Grandfather all believed we were not alone in this universe. That's probably why Dad spent so much time learning the constellations and the planets. That was why he watched the skies.

There is no way anyone could ever convince me that what I saw was a weather balloon, a flock of birds, or a kite. I have watched the skies on my own for years, but not with the dedication I had for doing it then. I don't have the time to do it like that either. I already believe we have shared our planet with other beings from another world.

One of these days, our government will reveal that they have been doing reverse engineering on vehicles they have had in captivity or many years. Can you imagine what would have happened if they had not kept this secret? Whoever held this vehicle held the proof of alien life and advanced technologies far beyond what we had.

Chapter Eleven

ABDUCTIONS AND MISSING TIME

How about the people who have had lost blocks of time and their only memory is what they doing before the lost time? Can there be such a thing as an alien abduction? Are all these reports fabrications of the abducted? Who's to say these same experiences weren't happening in ancient times, too?

Where's the evidence of theses visitations? If Area 51 is any indication of why we don't have any, I can see how that would happen time and time again. The evidence is scooped up and carried away from the public's eyes. In a recent survey, the people who believed that we are not alone was astonishing.

Interviews with possible witnesses have proven inconclusive many times. One of the most famous accounts of an alien abduction was the case of Barney and Betty Hill in New Hampshire in 1961. I remember the news story when it came out. It was frightening to me.

The Hills were willing to be put under hypnosis. The only thing that came out of this was that was odd was a map that was drawn, by Betty, of the stars where the aliens claimed they had come from.

   The map included information that was not available until a little later. The detailed description they gave of their actual abduction aboard the spaceship and what they went was through physically was disturbing. Is it true or just the hysteria of two people not able to rationally explain something that happened to them?

   Alien abductions are very much a reality to those who think they have experienced them. Implanting small devices under the skin, perhaps like a tracking device, have been reported for many years. Some of these devices have been removed according to the people who were abducted. What happened to these devices? Why haven't we heard any more about it? Did the government confiscate those devices?

   Another famous abduction was in one of the Carolinas in a lumber town. Supposedly, four men were out logging and just as they were leaving an event happened that has changed their lives forever.

   Three of the men had gotten into the truck to leave when the fourth man, who was standing in the forest looking upward, pointed out something in the sky. This bright light was moving in their direction and the others were in a hurry to leave and head back to town. The fourth man wanted to stay a minute to see what it was. The men began to drive off then circled back to get their friend. When they went back their friend was missing.

Many days later he was found, naked, visibly shaken and was in shock. He told an account of being beamed up into this ship, physically abused and probed, and could not remember how he got back to where they found him.

I believe a movie named "Fire in the Sky" was made about this incident. Of course, it was much exaggerated and embellished, but much of it was based on the facts as he recalled them. Was it a hoax just so it could be exploited and with the hope of lots of money to be made? They still stick to their story. Most of them still live there and work as loggers going back again and again to the spot where it happened. Maybe it did.

What about the animal mutilations we hear of in the news? The animals are surgically operated on and are left with perfect incisions as if done by a licensed physician. Why are they done and by whom? What are they doing with these parts and pieces? Is it to test their DNA and use it somehow? Is it for genetic testing or genetically altering the animal?

This has happened all over the world.

   Is our knowledge of cloning an accident or were we steered in that direction for a higher purpose? Are there hybrids made from these experiments with DNA? Were some of the ancient creatures pictured in the hieroglyphs in actuality hybrids? Are we hybrids?

   Modern day geneticists have learned how to duplicate body parts by growing them on live animals. I wonder how advanced we will be in another ten years? Will we be able to clone our own parts or organs needed because of being defective or destroyed accidentally? We are almost doing that now. Do you really think that some other country hasn't been duplicating an actual humanoid already? Not ever country is as scrupulous and conscientious as we are. It's no telling what has been happening behind closed doors.

Chapter Twelve

ARE WE TAKING 2012 SERIOUSLY?

    You can bet the farm there are many people taking this very seriously. Here in the United States there are old abandoned missile silos being converted to living compartments for a specified number of people. The ticket price of the units is very expensive. Not only are the builders providing "safe" units, they are, also, stocking up on survival food packets. Just notice what the prices for these previously cooked and preserved meals cost now if you can find them where they aren't on backorder. There's quite a market for this food.

    Each compartment is supplied with all the basic necessary requirements for survival. Most of them are being built with reading rooms and libraries. Movie theaters, pools, exercise rooms and equipment are on their own floor for everyone's use. It sounds as though they have been well thought out. Even the interior walls and exterior doors are designed to withstand a nuclear blast. Air ducts are hidden and protected from any attempts at vandalism. How do I know this? A news reporter was allowed to show the plans for one of these so they could help them be marketed. The exact location is never revealed.

There are people converting mountain caves into safe places. They are stockpiling food, water and arms and ammunition. All the participants are being taught how to shoot and are sharing survival techniques. This was a family project. The price of admission is contributing to the supplies for the family. It's sort of a poor man's version of the silo project.

In Europe, in a secret location, are buildings being constructed that are shaped like domes on top with the remainder going underground. These are designed to withstand a nuclear blast and are built high on a mountain to be able to stay dry in case there is a huge flood or tsunami. I think I remember one being in Spain. Needless to say, they are, also, very expensive.

We are all aware of the places around the United States that are for our President, the Cabinet, and so on, in the event of war or any other catastrophic event. Those have existed for a very long time, probably since the Cold War. One is said to connect at the airport in Denver where Air Force One can land and its passengers can go by tunnels to the safe place. Can you imagine what the tax payer dollar is for that to be built and maintained? One of the reasons they have three or four of these facilities is if one area gets hit harder than the other, then they are able to go to another one. It makes sense. Can you envision what chaos would be happening outside?

The idea of large unsinkable boats being built, similar to the ones in the movie "2012", are for those who think the "end" will be more like an asteroid assault or a huge Super volcano that causes the waters to rise to unfathomable heights. It wouldn't protect them from a nuclear attack though. I guess you have to pick your poison.

In Hutchinson, Kansas, deep in a salt mine covering about thirty-five football fields, is a storage facility that contains many important documents, records, paintings, works of art, books, and anything that could be used for reference and guidance in case of the event of a disaster or total destruction of the outside layer of the earth. The climate down in the salt mine is perfect for this type of storage.

In Norway, there is a facility housing twenty million seeds of food plants. It has to be manned and the stored seeds have to be constantly replaced and rotated so that only the freshest seeds would be available for planting.

In another location, which wasn't disclosed in the report as to where it was, is the DNA of every animal on earth. It is a giant cryogenic storage bank. I'm not sure what other DNA is there. Who knows, maybe different examples of human DNA in case we get totally obliterated.

There is no telling how many of these banks are out there and where they are. Certainly other countries are doing the same thing we are. Preparations for starting over" have to be considered.

What does the average person do to survive? Not much, I'll bet. Unless you're rich, I don't think you stand a chance of getting into one of those "safe" places. I'm not sure I would want to anyway. I'll just have to pray and take whatever comes my way!

Let's hope it isn't the end, but just a warning for us To start o car bout wht we do to each other an how we treat our planet.

Chapter Thirteen

ASSIGNMENT: COLONIZATION

In so many of the ancient civilizations, the worshipping of many gods was regularly practiced. What if your agree there may be other intelligent life out there? What if there were many different beings from our Milky Way or beyond? Then, what if Planet Earth was part of an experiment that was assigned to several extraterrestrial "gods"? What if we were merely a contest among aliens from different planets to see who could make the most capable, most advanced civilization in a certain amount of time? An assignment like homework. Or maybe for their own amusement.

There may have been basic rules to follow using the same beginning creatures. What if there were many Adams and Eves all over the continent, each developing at their own pace? Each maybe brought to an area by a different alien. What if only a certain amount of alien visitations were allowed to give their humans a little help when they were struggling? What if they could only do certain procedures like artificial insemination or genetic reconfiguring so many times to affect the evolution of man? What if they had their choice of picking certain land animals and sea creatures?

When we didn't perform to their expectations, did they punish us with disease, famine or wars? Did they totally annihilate some civilizations and then have to quit the game and go home after the annihilation only to wait until they could come again?

This theory may be bizarre but, it would explain why we had advancements far beyond what we should have been capable of having at any particular period. It would explain the total disappearance of missing civilizations. The floods and all other natural catastrophes could be explained as punishments or as challenges for us to overcome. It could have been part of the game.

The weird creatures pictured on the walls, the hieroglyphs, and drawings could have been that way because they were from different planets. The civilizations were painting what they saw.

Maybe the tools and equipment needed to build the wonders of the world were only allowed to be used to build them, but not kept. That would certainly explain why there were no intricate tools found that would be capable of constructing these incredible masterpieces. The rules allowed them to keep only basic tools.

What if they would earn points or credits for different

levels reached by their particular society and as a reward were given other levels of intelligence to pass on to them when they earned these credits?

What if 2012 was the end date of the games? What if the winner would be rewarded with a new Utopian-like world to live on with his choice of who would be picked to stay? Then another Adam and Eve would once again be chosen to sire the beginning of the new world of hybrids chosen by the god. Maybe the Planet Earth is the prize itself.

What if all the sightings we have had are the extraterrestrials checking in on their project? These ETs are obviously extremely more capable and intelligent than we are. Their technologies are probably beyond anything we could even begin to imagine.

What if there was a more malevolent reason for their visits? What if "man" had been made to serve the aliens like to mine the valuable ores and minerals our Earth has so abundantly? Or to manufacture the gold they may have needed from the pyramid? Gold seems to be one of the most valuable of the metals and is prevalent in every civilization.

Even more frightening, what if we were bred to be a food source? Could that be why entire civilizations have disappeared without a trace? Do we get gathered up and transported only to be eaten at some point? Are we getting ready to be thinned out again because of the over population we are experiencing? Are we going to be "soylent green" like the movie suggested? Just maybe they decided if they tried to eat the animals on the Earth, it would be a lot simpler to consume humans because they would be take care of themselves. Is this so absurd to think about?

Why would we have alien visitations at all unless we are an experiment of some kind or because the Earth has so many rich minerals and metals to mine? Maybe their world's supplies have run out so they have to look elsewhere. Maybe they just need ours, but they can't survive in our atmosphere.

If they wanted to invade Earth, don't you think that would have already happened? Certainly, their technologies would dwarf ours. They must be extremely advanced. I somehow don't think that would be the reason.

Remember, this is not my personal opinion, but just another theory to throw out there. After all, if you were trying to explain genetics to a caveman, how far fetched would that sound? How impossible that would sound.

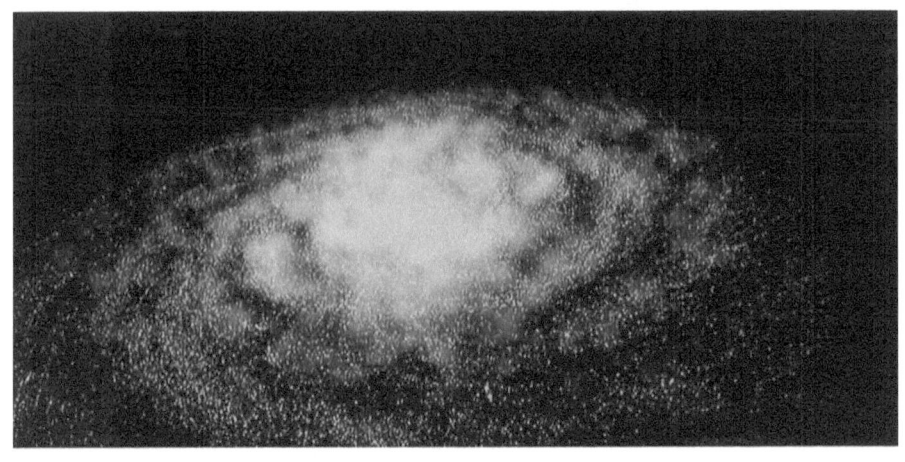

Chapter Fourteen

WHAT DOES ALL THIS MEAN?

If you are a believer in extraterrestrials or if your not it doesn't really matter. That's your prerogative not to believe if you so choose. But let me point out that you may be all by yourself in this opinion. Mainstream scientists would beg to differ with you. More people believe in extraterrestrial life now than ever before.

Our ability to send satellites far into our solar system and send back pictures and samples is turning up information that makes it more probable every day that its real. It's just a matter of time before we meet ET. Some of us may already have done that.

Will they be friendly or will they seek to destroy us and take over our world? Or will we find out they have been here many times before? Will we be like the primitive cavemen to them or will we find out that they have been engineering our progress all along to a point that would be useful to them? Will we find out they have visited us for many centuries and been mining our precious minerals for use on their planet Because these have been depleted or destroyed on their planet?

The world's population is exploding at an alarming rate. I believe I saw a chart recently that said 57% of the world's population are Asian. Are we getting to the tipping point? Do we need to be "thinned" out some?

Did the Mayans leave us a warning that they will be back in 2012? Was the prediction of Nostradamus letting us know that we have something or someone coming from the heavens that will end our world as we know it? Can they all be wrong? I don't think so.

Doesn't the Book of Revelations in the Bible forecast the eventual annihilation of the species we call man? Only the "chosen" will be saved to continue on and live in Paradise. Are we going to be cherry-picked so that only the best will be saved to go forth and multiply until the tipping point is reached again?

The Mayan calendar is based on what they called ages. It is not a prediction for the end of all times for all of us. It's a new beginning for those who can survive. Just like the Bible says we will have a new beginning. The word Revelation means a revealing or unveiling. Are we soon to have the ultimate truth revealed to us? Only time will tell.

I choose to believe that whoever or whatever this God is who is promised to come, will let us continue on like he has promised. I believe He is a Saviour. Who will survive after all the scourges have been released upon us will be unfolded soon if you are true to your religion and believe.

Has the Earth we live on just been like an assignment for one of many extraterrestrials that may have colonized us and the year 2012 is when the assignment ends? Or is it the end of all life on Earth due to the total destruction by an asteroid, nuclear war, a virus we can't treat or control or is it the combination of all these things? Will it be repopulated with hybrid beings brought to Earth from the visitors from the stars? I'd like to believe that's not going to be the case.

What can we do to prepare for this eventual apocalypse? Go on living the way you are now, but try to live with integrity and show you care for your fellow man. Help a neighbor or a stranger when you can.

Most of us won't have the money to be able to go to a safe place. That may not make any difference anyway.

If we have nuclear wars throughout the world, the air and soil will be polluted for many years. If you even can survive the initial blasts.

If we are going to be hit by some huge celestial event like an asteroid, then the Earth will be covered by the devastation that goes with that. Similar to a super volcanic earthquake, there could be tsunamis, floods, storms, ash over everything, fires, no sunlight and dark skies.

If the event was that we had a polar axis shift, now that we know we had one only 5200 years ago and it is possible, there would probably be total destruction of the whole world. Nothing would survive.

So what do think you should be doing? How about going on just like you do normally? Let's hope none of these predictions will come true. Let's hope they are all wrong. After all, it wouldn't be the first time we've heard about the end of the world.

www.ingramcontent.com/pod-product-compliance
Lightning Source LLC
Chambersburg PA
CBHW030905180526
45163CB00004B/1710